MYSTERIES OF THE UNIVERSE

MYSTERIES OF THE UNIVERSE SERIES

MYSTERIES OF THE UNIVERSE

Franklyn M. Branley

Diagrams by Sally J. Bensusen

LODESTAR BOOKS   E. P. DUTTON   NEW YORK

LIBRARY OF CONGRESS CATALOGING IN PUBLICATION DATA

Branley, Franklyn Mansfield, date
   Mysteries of the universe. ·

   (Mysteries of the universe series)
   "Lodestar books."
   Bibliography: p.
   Includes index.
   Summary: Discusses the various theories concerning
the creation, expansion, and possible end of the universe.
Also defines such phenomena as black holes, neutrons,
pulsars, red shifts, and other discoveries that further
enhance our knowledge of the world beyond our planet.
   1. Cosmology—Juvenile literature.   [1. Cosmology.
2. Universe]   I. Bensusen, Sally J., ill.   II. Title.
III. Branley, Franklyn Mansfield, date.   Mysteries
of the universe series.
QB981.B734   1984   523   83-25302
ISBN 0-525-66914-0

Published in the United States by E. P. Dutton, Inc.,
2 Park Avenue, New York, N.Y. 10016

Published simultaneously in Canada by
Fitzhenry & Whiteside Limited, Toronto

Editor: Virginia Buckley   Designer: Riki Levinson
Printed in the U.S.A.   W   First Edition
10 9 8 7 6 5 4 3 2 1

Photograph on opposite page
courtesy of Palomar Observatory

CONTENTS

MYSTERIES OF THE UNIVERSE

1 THE NUMBER ONE MYSTERY

Our senses tell us about our surroundings. We can touch our planet, see it, smell it, and hear its sounds. We have many ways of knowing that it exists. But this is not so with the Moon, the planets, the Sun, and the stars. We can see the Sun and feel its warmth. We can see the Moon and a few of the planets—Mercury, Venus, Mars, Jupiter, and Saturn. With telescopes we can also see Uranus, Neptune, and Pluto. With our eyes alone, we can see thousands of stars. Millions can be seen with a telescope. But objects we can see even with the most powerful telescopes make up only a small part of the universe.

The word *universe* means all together, the whole thing, everything turned into one. We could say the universe is all we can see or otherwise know about. That means everything we can learn by using special instruments to capture radio waves, X rays, and other kinds of radiation that come from space. Down through the ages, people have been challenged to build better instruments to learn more about the universe and chances are that the search will go on and on.

Ancient people had no instruments for looking into space.

They had only their eyes. With them they could see the Sun, the Moon, five planets, and a few thousand stars, just as you and I can. From their point of view, that was the entire universe.

In the beginning of the seventeenth century, people looked into the sky through low-power telescopes. They saw that there were thousands more stars than they had known about. The view of the universe through a telescope was quite different from that seen with the eyes alone. Suddenly the universe became greater by far.

During the centuries following 1610—the year when Galileo, the Italian astronomer, became the first person to use a telescope for stargazing—better and better telescopes were built. They improved the picture of the universe. But right up to the 1920s, people still believed the entire universe was contained within our own stellar neighborhood—to them the boundaries of the universe, if any, were marked by the outer stars.

Early in the 1920s, astronomers became interested in a hazy patch of light they had detected in the sky. They studied the area carefully and discovered that the patch of light was made of separate stars. There were millions of them. The patch was actually another stellar neighborhood far beyond our own—2 million light-years away. It was another galaxy.

Our galaxy, the Milky Way, is a giant formation 100 000 light-years across—it takes light 100 000 years to travel from edge to edge. Each second light goes 300 000 kilometers—so the distance across our galaxy is enormous. The galaxy is shaped somewhat like two fried eggs back to back, thicker at the center than along the edges. It is made of the Sun, the Moon, and the planets that were known to the ancients. It is also made of vast amounts of dust and gases. But, for the most part, it is stars, perhaps as many as 200 billion of them.

Nebulas such as those on pages 6–8 are formations of gases within our own galaxy. In many of them, the gases are packing together and stars are being born, as can be seen by the small, dark spots.

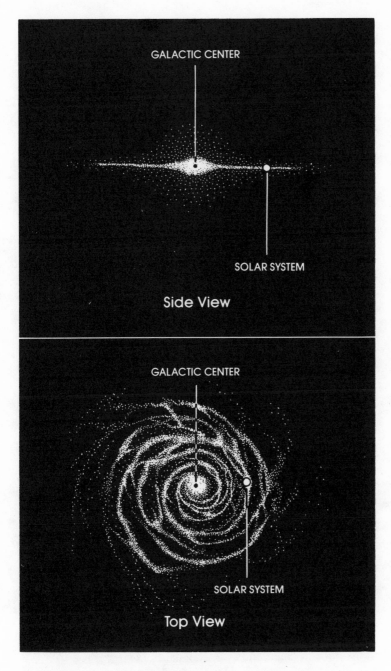

GALACTIC CENTER

SOLAR SYSTEM

Side View

GALACTIC CENTER

SOLAR SYSTEM

Top View

Our galaxy is one of billions that makes up the universe. Scientists believe it looks somewhat like this, and we are located some 30 000 light-years from the central portion. The entire galaxy rotates, taking 250 000 000 years to complete one turn.

Lagoon Nebula in Sagittarius. PALOMAR OBSERVATORY PHOTO-GRAPH

Great Nebula in Orion.   MOUNT WILSON AND LAS CAMPANAS OBSER-
VATORIES, CARNEGIE INSTITUTION OF WASHINGTON

Trifid Nebula in Sagittarius.   PALOMAR OBSERVATORY PHOTOGRAPH

In the next few years, thousands of other galaxies were discovered—collections of billions of stars at distances from us and from each other measured in trillions of kilometers. No one knows how many galaxies there are in the universe. Certainly there are billions of them.

In just a few years the universe became much vaster than anyone had suspected; from a measurable universe it had become practically without end. It is believed the universe spreads out some 26 billion light-years. There are some 10 trillion kilometers in a light-year—the distance light travels in a year. No wonder the universe remains a major mystery, for who can comprehend such distances?

Within our universe there are endless mysteries, some of which will be explored in the next chapters. The greatest of all mysteries about the universe may be that the brains of human beings can comprehend it. Or can they? Perhaps we'll never be able to understand entirely the complexity of the universe—the way it works, how it all fits together.

However, scientists think they know something about where it came from, how old it is, and what it is made of. Such things will be examined in the pages that follow. You'll find what has been discovered and what continues to challenge people who explore the universe.

2 THE START OF IT ALL

How old is the universe?

Your body is made of billions of molecules. Many of them contain hydrogen, which is the simplest of all substances. The core, or nucleus, of hydrogen contains one proton—a massive particle that carries a single charge of electricity. Since it has one proton, the atomic number of hydrogen is 1. The helium nucleus contains two protons, so its atomic number is 2. And so it goes—uranium, for instance has ninety-two protons in the nucleus, and its atomic number is 92.

The hydrogen in your body may have been formed at the time the universe began. That would be some 15 billion years ago. Many astronomers believe this is the age of the universe. Background temperature, which is discussed below, supports the idea. Also, the outer limit of the universe seems to be around 15 billion light-years away. Suppose something, such as a star, is 10 light-years from us. That means it took light 10 years to travel from the star to us. Therefore, we are seeing the star as it was 10 years ago. If the limit of our seeing were 15 billion light-years, the star would have released the light 15 billion years ago. The objects farthest from us seem to be around 13 billion light-years away—so they are 13 billion years old. The universe itself would be a bit older, for it would have taken an additional 1 or 2 billion years for the objects to have formed.

Hydrogen was the first substance to be created. Eventually other elements were made, using hydrogen as the building material. Now there are countless numbers of substances in the universe, but you and all the people of Earth, everything on Earth, Earth itself, everything we see in the sky and vast areas of things unseen, had their start at the same time. It was some 15 billion years ago—time zero, the time of the big bang.

How did the universe begin?

No one knows for certain—it is a major mystery. But there are theories. One is called the big bang. The name was coined by George Gamow, who was developing an idea first suggested by Georges Lemaître, a Belgian astronomer, in the 1920s.

Time zero was the beginning of time and space. Before that, matter involved in the big bang existed, though probably in the form of energy that was spread out like a thin gas. All this material was to become the primeval or primitive atom. It has also been called the cosmic egg.

No one is sure where the material came from. Some have suggested it may have come from still another universe, one that existed long before the one we presently know about. That may be so. But we have no way of knowing for sure whether or not there ever was a pre-universe.

It is quite certain that at time zero no atoms existed. Atoms, you remember, are combinations of several particles. For example, an oxygen atom is made of eight protons, eight electrons, and eight neutrons—massive particles that carry no electrical charge. Temperature at the beginning was very high, so high that it would have prevented the particles that did exist from combining to form atoms. So there was no oxygen, nitrogen, calcium, or any other material. There were bundles of energy called photons—massless particles of light, X rays and other forms of energy. Also, there were numerous kinds of subatomic particles. That means there were parts of atoms, but they were not yet fully combined and holding together. They combined, but separated just as fast as they formed, so they were in balance.

The particles were packed together more closely than can be imagined; density was unbelievably high. The density of water is 1, meaning that 1 cubic centimeter of water weighs 1 gram. The density of the cosmic egg, the concentration of the matter of the universe, was incredible—in the order of 1 quadrillion—meaning 1 cubic centimeter weighed 1 quadrillion grams. Because of the intense packing of this primeval egg, scientists have computed that temperature would have reached at least 1 trillion degrees. No one knows for sure the kinds of reactions that might take place under such unbelievable conditions of density and temperature.

However, it is believed that the cosmic egg could not have held together. As temperature went up, so did pressure. It became so great that the cosmic egg exploded, causing it to expand rapidly. During the first few fractions of a second, density dropped a thousandfold, as did temperature. Everything moved away from the central point and is still moving today—the process of expansion continues. Moments after time zero, photons collided and combined. Units of energy were converted into mass. Protons were created—the cores of hydrogen atoms. Also, neutrons were formed—the massive parts of present-day atoms.

Most of the protons flew away from the center of the reaction. They are the hydrogen cores of the universe. Occasionally a proton combined with a neutron, or even with two neutrons. The particles formed were deuterons (two parts) and tritons (three parts). About one-fourth of the protons combined with other protons, making helium with atomic number 2—two protons. That helium still exists today, comprising one-fourth of the universe. Its presence in that amount is strong support for the entire big bang theory.

All of these particles—photons, subatomic particles, protons, neutrons, tritons—were moving away from the explosive center, many at the speed of light—300 000 kilometers a second. Many may still be moving at that speed. If they are, though, we'll never see them, for their light would be unable to reach us.

Not all the particles continued to move at high speed. Some collided with others, slowing them and cooling them enough so they held together. Protons built up more complex substances, those with higher atomic numbers, such as magnesium, aluminum, carbon, oxygen. They built all the way to iron, which is atomic number 26. These are the "old" elements.

Those with atomic numbers larger than 26 are younger. These materials were formed during star explosions, many of which occurred 2 or 3 billion years after time zero. Star explosions still occur today and, just as long ago, new atoms are still being produced.

All the new and ancient atoms add up to a very small part of the universe. By mass, 76 percent of the universe is hydrogen, and 23 percent of it is helium. All the other substances—iron, lead, gold, oxygen, neon, and so on—comprise only 1 percent.

How and when did galaxies form?

If there was a big bang, there should still be expansion. And there is. All parts of the universe seem to be moving farther apart. At the same time, gravity is working to pull all the masses together—to slow down the expansion.

Perhaps a billion years after the big bang, expansion and temperature had dropped enough for eddies to form in the expanding material. Small clusters of particles formed. Such clusters were more massive than the surrounding material, and so they exerted a greater attraction. More and more material was pulled into the clusters—in our galaxy, the material became at least a billion times the mass of the Sun.

At the same time, other masses of equal size were forming, and these also were to become galaxies. Our new-formed galaxy pulled on neighbor galaxies, and we were pulled by them. The huge masses were set spinning. This spinning caused the mass, which had been spherical, to flatten gradually. Our galaxy became a disk, bulged at the center and thinning out toward the edges.

Gases composing galaxies were in constant motion, and they were pulled into orbits around the central region. Here and there atoms clustered together and built up masses great enough to become stars. In the case of the Sun, it is believed that happened some 5 billion years ago. Since the galaxy may be some 12 or 13 billion years old, our Sun is a "young" star.

Planets are leftovers. They are made of some of the materials that were not gathered into stars. In the scheme of the galaxy, and of the universe, planets are of minor importance. For example, in our own solar system all the planets together add up to less than 1 percent of its mass. Just about all the remaining material is in the Sun itself.

What is the inflationary universe?

A good many astronomers believe that the story of the universe is told quite accurately by the big bang theory. But not everyone. A few astronomers think that the universe is too much alike throughout to have been formed by an explosion. They think there would have to be greater differences than are observed. They feel the expansion is too smooth. Also, it is strange that the rate of expansion is just right for galaxy formation. If it had been faster, galaxies would not have formed; and if it had been slower, the universe would have collapsed.

These astronomers are suggesting that the universe began

from nothing, or almost nothing. There was energy, and somehow this energy became concentrated and was converted into additional mass. There was tremendously rapid inflation—billions of times greater than we know about. The known universe condensed inside that great inflationary volume.

They believe there is a superlarge universe beyond our vision. Within that super universe there are regions such as ours—the billions of galaxies. But these regions add up to very little. Most of their universe lies far beyond the "seeable" boundaries.

The theory would be another way of explaining the expanding universe, and the background temperature of the universe. Astronomers who support the theory think its explanations are better than those offered by the big bang theory. Therein lies another major mystery.

Is the universe expanding?

The big bang theory implies the universe should be expanding just as observations indicate. Galaxies are moving away from one another. We are moving farther from neighbor galaxies, and they are moving farther from us. The movement is much like that of dots on a balloon as it is filled with air; all dots move farther from one another. The universe is expanding.

Astronomers know this because they can measure the red shift, as discussed in Chapter 4. The red shift means that if a star or galaxy is moving away from you, its light is a bit redder than it would be if the distance were unchanging. This reddening of light is called the red shift. The greater the movement, the greater the red shift. The light of galaxies shows red shifts. And the most distant galaxies have the greatest red shift. That is exactly what astronomers expected; the farthest galaxies should be moving the fastest. That is because they are made of the outer portion of the cosmic egg, the part that would be expected to have the highest speed.

What is the temperature of the universe?

If there had been a superhot cosmic egg some 15 billion years ago, its temperature would have dropped sharply from that hot beginning of a trillion degrees and more. Astronomers can figure out that by now the temperature should have dropped to three degrees. And that's what it is. This is three degrees (3 K) in the Kelvin or absolute scale. In the Celsius scale, the coldest anything can get is −273° C. Zero K is the same as −273° C. So 3 K is three degrees above the coldest anything can ever get— the absence of all heat.

In 1965, Arno Penzias and Robert Wilson, scientists at the Bell Telephone Labs in New Jersey, discovered what is called cosmic background radiation. Wherever they looked between the stars and galaxies, they found lingering radiation. And they found that it measured very close to 3 K, just as astronomers had said it should. The discovery was important, for it was strong proof that the universe began and evolved as theory had predicted.

In 1965, Robert Wilson (left) and Arno Penzias (right) discovered that the temperature of the universe is three degrees. It is the temperature to be expected if there had been a big bang some 15 billion years ago. COURTESY OF BELL LABORATORIES

In spite of great strides such as this that have been made in understanding the universe, mysteries still abound. Will the universe expand forever, or will it stop and then begin to collapse? Can we be sure about how it all began? Did the universe "inflate," or was there a big bang?

We may never find the answers. But we shall certainly get closer to understanding it all.

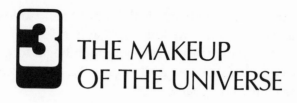

THE MAKEUP
OF THE UNIVERSE

Why is there so much hydrogen?

In the first chapter it is said that hydrogen is the most abundant substance in the universe. Also, you remember, it is the simplest of all the elements, made up of one proton (which carries a positive charge of electricity) and one electron (a negative charge). The next most abundant element would be the one that contains two protons, which is helium.

Hydrogen and helium are the oldest elements in the universe. They were formed in the first few moments after the big bang. This is what scientists believe from their study of the structure of atoms and the energy needed to produce them. A good many of the elements were produced only minutes after the explosion—but hydrogen and helium were formed within fractions of the first second.

Immediately after the big bang, you recall, energy combined to produce matter. At once, most of the matter broke down again to produce energy. There was great instability—conditions changed radically and rapidly. After only a hundredth of a second or so, conditions settled down somewhat. Protons were the principal form of matter that was created—and these were the nuclei of hydrogen atoms.

In a good many cases, protons combined with neutrons. Often they did not hold together. But, when they did, heavy hydrogen was formed. The core of this substance is a deuteron—a proton and a neutron together. When two neutrons combined with a proton, a triton was made. And the substance was called tritium.

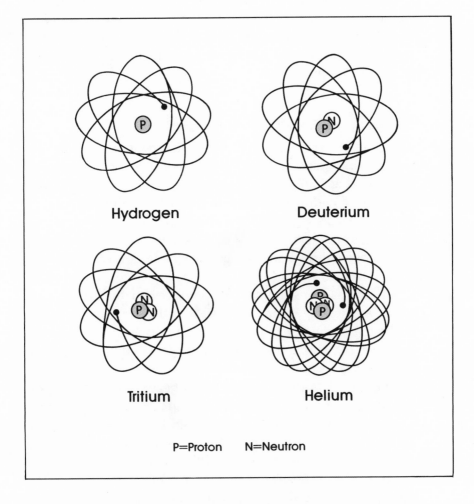

Hydrogen   Deuterium

Tritium   Helium

P=Proton   N=Neutron

Hydrogen is the most abundant substance in the universe. The basic building block of hydrogen and the other elements is the proton. A proton and a neutron make a deuteron—deuterium when an electron is added. A proton plus two neutrons make a triton—tritium when an electron is added. Two protons and two neutrons make a helium nucleus.   **19**

Protons are the building blocks of the universe. They combine with negative electric charges to produce neutrons. And by various combinations of protons, and protons and neutrons, other elements are built. For example, during the first few minutes of the universe, deuterons combined to form helium. These were fusion reactions, and during them energy was released. In turn, the energy caused additional fusion reactions to occur, and so many of the lighter elements came into existence. However, reactions beyond helium must have occurred only rarely, for 99 percent of all the atoms in the universe are hydrogen and helium.

Why are heavier elements rare?

The chart on pages 22 and 23 shows that hydrogen and helium are very abundant, and the other elements are rare. Milleniums after the big bang when galaxies formed, they were mostly made up of hydrogen, as were the first stars within the galaxies. The oldest stars we have been able to observe are made of hydrogen. Even newer stars, such as the Sun, contain a great deal of hydrogen, though the Sun has larger amounts of heavier elements than do the ancient stars.

Fusion reactions are self-sustaining when the reaction releases more energy than is needed to cause the reaction. When hydrogen nuclei fuse to form helium, photons (units of energy) are given off. This energy is enough to cause additional protons to fuse and become helium. So the reaction continues.

Other nuclei fuse to produce even heavier elements. For example, right now the Sun is converting hydrogen to helium. Eventually it may become hot enough to convert helium into carbon. But chances are that the fusion process will end there—the Sun does not have sufficient mass to continue beyond. However, heavier elements were made early in our history, and they are still being made today.

Many ancient stars were extremely massive. Their gravitation was so great that they packed together tightly enough to generate tremendous heat. There was enough to cause them to blow up. When they did, there was sufficient energy to force "unusual" fusion reactions—the kind that produced the heavier elements. The superstars of ancient days are sometimes duplicated today, for occasionally there are star explosions. The energy they give off is incredible. During the explosion as much as half the material of a star may be thrown into space.

What are supernovas?

Originally hydrogen and helium were the only substances in the universe. There may have been occasional occurrences of a few other elements, but they were extremely rare.

Several billion years after the big bang, galaxies formed. And with the galaxies, stars came into existence. They were hydrogen stars. That means they were massive enough and hot enough for fusion reactions to occur inside of them. Protons combined to form helium.

As helium builds up, the star temperature increases. It becomes hot enough for helium to fuse together in a series of reactions that produces carbon and oxygen—and more energy.

Stars a lot more massive than the Sun, and a lot hotter, are carbon stars. In them, carbon is the raw material. Fusion reactions in such stars produce oxygen, neon, sodium, magnesium.

As mass increases, so does temperature. That enables other reactions to occur which produce elements with even higher atomic numbers. These are the temperatures and masses of some massive stars and the reactions that occur within them:

TEMPERATURE KELVIN DEGREES	MASS (TIMES MASS OF SUN)	CHANGE
1 billion	5–10	Neon becomes oxygen, magnesium.
2 billion	10–20	Oxygen becomes elements from magnesium to sulfur.
3 billion	20 and more	Silicon produces elements up to iron—number 26.

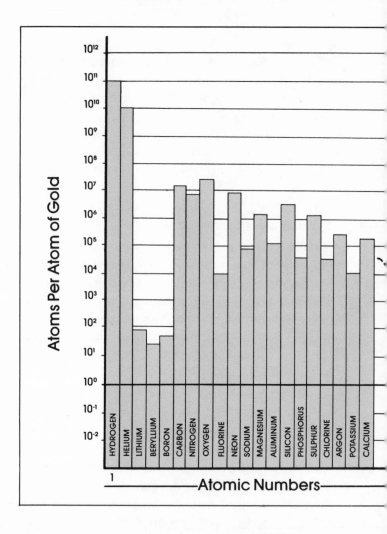

For the most part, the elements up to iron (number 26) are a million or more times more abundant than the elements with numbers greater than 26.

Fusion reactions stop at iron because it takes more energy to fuse iron than is released during the reaction. Therefore a star could not sustain the reaction—the star would be cooled by it instead.

How then do we get the elements with atomic numbers greater than 26? They come from exploding stars, called super-

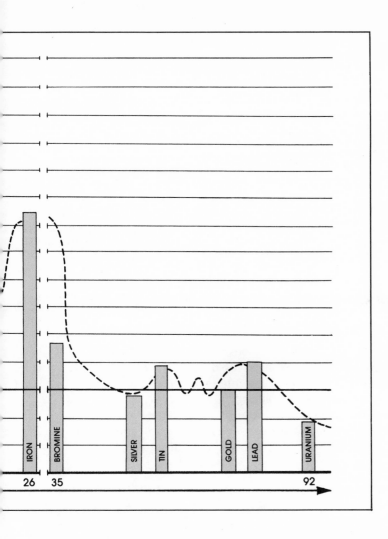

novas, some of which may be sixty or more times more massive than the Sun. They explode with tremendous force, creating temperatures of trillions of degrees. And that's enough to force heavy elements to fuse and so make even heavier ones—all the way up to uranium and even beyond.

Fusion reactions may go on inside a star, building up the elements until it becomes an iron star. Then fusion can no longer occur, so the star cools. But not for long, because as it cools, the star contracts; it shrinks and packs together. This causes the temperature to go up again.

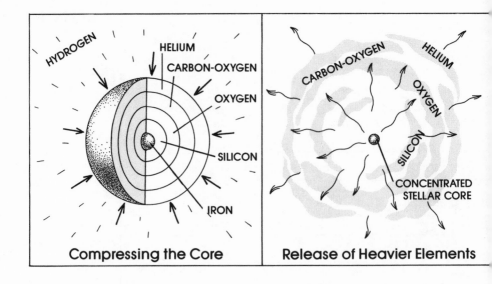

HYDROGEN
HELIUM
CARBON-OXYGEN
OXYGEN
SILICON
IRON

Compressing the Core

CARBON-OXYGEN
HELIUM
OXYGEN
SILICON
CONCENTRATED STELLAR CORE

Release of Heavier Elements

In a massive star, the core may be iron. Fusion reactions outside the core have produced layers of other elements as shown. Packing of material makes the interior temperature soar. Iron breaks down into helium, causing the star to collapse. This heats up the central area, causing fusions which release enough energy to blow the star apart. Pressure inward packs the material to make a superdense core, a special kind of star.

The iron core of such a star may be surrounded by a layer of silicon, and around that is a layer of oxygen. Both layers become hot enough (because of the contraction) for fusion to occur in them. The center of the star heats up too. When temperature reaches 5 billion degrees, protons have enough energy to enter the iron nucleus. They cause the iron nucleus to break down into helium. Tremendous heat is used up in the process. That means rapid cooling, and further collapse. Once more, temperature zooms upward and the protons and neutrons set free from the iron move fast enough to enter the cores of atoms. They are changed to heavier elements.

Billions of fusion reactions occur—heat is consumed, the star collapses and a tremendous blast is created. Often it is equal to the explosion of billions of atom bombs. The star blows apart, and all kinds of elements are blown into space. They are stardust. Other stars may be built from them. Also, many of the elements eventually fall to Earth and are taken up into our own bodies through the air we breathe and in our food and water. Each one of us is made of stardust.

What is technetium?

Technetium is element number 43, and its story is further proof that the elements are made in the stars.

By studying starlight, astronomers are able to determine the substances in a star. Each material gives off its own peculiar light. Along with other elements, technetium has been found in several stars.

Technetium is an unstable element, meaning that it gives off particles and so breaks down into other elements. And it does so rapidly. We say the half-life of technetium is 220 000 years. In that many years only half the original amount would remain; in 440 000 years there would be one-fourth the original amount, and so on. After 5 million years, which is only a short time in the life of a star, only a billionth would remain. So, if technetium is identified in a star, the technetium must be "new," for it would have to be younger than the star itself, which may be several billion years old. The technetium had to be created within the star.

The discovery of technetium was an exciting event, for it was definite proof of the theory of element formation in the stars.

Scientists have found part of the solution to the mystery of how the elements came to be. As frequently happens, these answers have led to other mysteries. For example, after a star explodes, blowing half of itself into space, the core of the star remains. What happens to it?

That's another mystery, one explored in Chapter 5. Before that, let's see how scientists are able to determine the size of the universe.

Until early in this century, galaxies such as this one were believed to be nebulas within our own galaxy. Now we know they are far beyond it. This one, the great galaxy in Andromeda, is some 2 000 000 light-years away. PALOMAR OBSERVATORY PHOTO-

THE SIZE
OF THE UNIVERSE

No one really knows the size of the universe. The best we can do is get some idea of it. We are limited to what we can observe—the part we can see or learn about by light waves, radio waves, X rays, or some other form of radiation. Other than that, we must speculate—depend upon therories.

How close are neighbor worlds?

Outside our own galaxy there are billions of other galaxies, most of them at great distances. The two galaxies nearest to us are called the Magellanic Clouds. They can be seen from the southern hemisphere, and were first reported by Ferdinand Magellan when his expedition made the pioneering voyage around the world in 1519. The Magellanic Clouds are 160 000 and 190 000 light-years away.

The Andromeda galaxy, shown opposite, is one that we can see from the northern half of the world as a hazy patch of light in the autumn skies. It is also closer than most other galaxies, even though its distance is 2 100 000 light-years away. This

Barred spiral galaxy in Eridanus. PALOMAR OBSERVATORY PHOTO-
GRAPH

Unusual type of galaxy in Centaurus. PALOMAR OBSERVATORY
PHOTOGRAPH

Spiral galaxy in Triangulum.   PALOMAR OBSERVATORY PHOTOGRAPH

Cluster of galaxies in Corona Borealis, about 1.3 billion light-years away.   PALOMAR OBSERVATORY PHOTOGRAPH

means that in kilometers its distance is 2 100 000 times 10 trillion—the distance that light travels in a year. Such incredible distances are hard to understand. Yet, as was said, Andromeda is a nearby galaxy. It is one of about twenty which altogether make up what is called the local group.

What are light-minutes, light-hours?

A light-year, as you probably already know, does not measure time; it is a measure of distance. Astronomers also use light-minutes and light-hours to measure distances. For example, the Sun is 8 light-minutes away—it takes sunlight about 8 minutes to reach us. The edge of the solar system is 5½ light-hours away. This means that when we look at Neptune and Pluto, the most distant planets in the solar system, we see them as they were 5½ hours ago. When we look at the Andromeda galaxy, we see it as the galaxy was 2 100 000 years ago. We are receiving fossil light.

Astronomers have measured distance far beyond the Andromeda galaxy. They believe that the most distant galaxies are some 13 billion light-years away. When we see them, we are receiving light that is 13 billion years old—and we see the galaxies as they were that long ago.

What are red shifts?

How can anyone be sure about such great distances? How do you measure them?

Light from a distant galaxy is picked up by a telescope and led into a spectroscope. That's an instrument that separates light into the wavelengths—the spectrum of light—of which it is made. If the light were that of a bulb, the spectroscope would show that the light is actually made of several wavelengths. They show up on a spectrograph as bright lines ranging from red lines (the longest wavelengths) to blue lines (the shortest). Our eyes see all the wavelengths or lines blended together; the spectroscope "sees" each separate line.

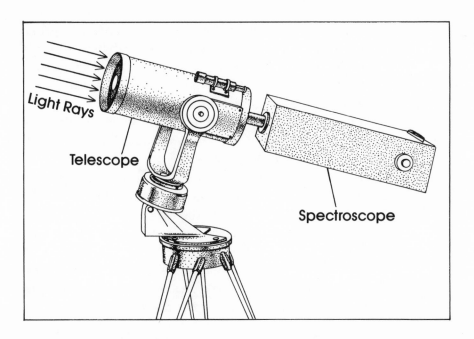

A spectroscope produces a spectrum of lines made by the various wavelengths of the elements that make the light.

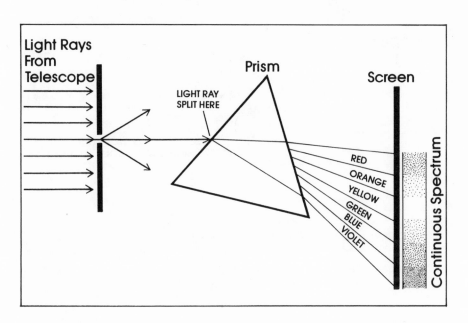

Each element gives off distinct lines, its "signature." When hydrogen light, for example, is produced in a lab here on Earth, the spectroscope always sees the lines at certain definite locations. If the source of the light were moving toward the spectroscope, all the lines would be shifted toward the short-wave end of the spectrum. This would be toward the blue light—the light source would appear bluer than it would if it were standing still. It would be called a blue shift.

Suppose we were aiming a spectroscope at a star and we found that the hydrogen lines were all moved a bit toward the long-wave end of the spectrum—slightly toward the red. There would be a red shift, meaning that the star is moving away from us.

Not only does the shift tell us whether a galaxy is moving toward us or away from us; it also indicates the speed of the motion. The greater the shift, the greater the speed.

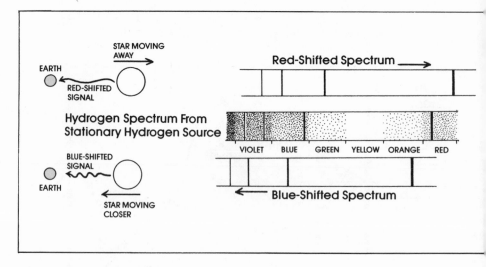

When hydrogen is at rest, its lines fall at the locations shown. When moving away, all the lines are shifted toward the red; the greater the speed, the greater the amount of shift. When moving closer, the lines are shifted toward the blue.

Relation Between Red Shift and Distance
for Extragalactic Nebulae

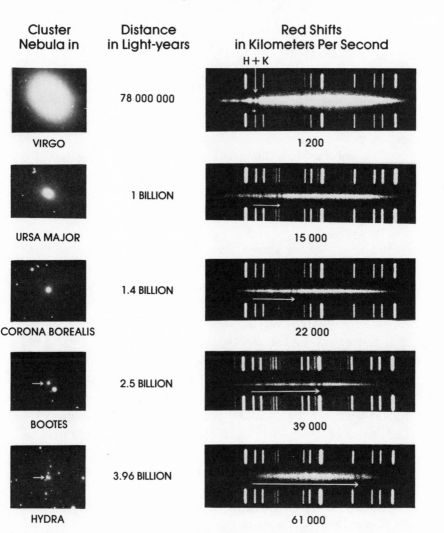

Cluster Nebula in	Distance in Light-years	Red Shifts in Kilometers Per Second
VIRGO	78 000 000	1 200
URSA MAJOR	1 BILLION	15 000
CORONA BOREALIS	1.4 BILLION	22 000
BOOTES	2.5 BILLION	39 000
HYDRA	3.96 BILLION	61 000

The farther away a galaxy is—the faster it is moving—the greater the red shift. Red shifts for two calcium lines (H and K) are expressed as kilometers per second. One light-year equals about 9.5 trillion kilometers. PALOMAR OBSERVATORY PHOTOGRAPH

Are red shifts related to distance?

Astronomers have been able to study certain separate stars in some of the other galaxies. And they know those stars are the same in many ways, such as composition and age, as some of the stars in our own galaxy. But they appear much dimmer. Their dimness can be measured. We know that light gets dimmer with distance, and we know how much it diminishes. Therefore, when we know the dimness of the star, we know its distance, and so the distance to the galaxy it is in.

Also, since we can determine the red shift of the galaxy, we can see how distance and red shift are related. We now have a method of finding the size of the observable universe—the part that we can see—a way to make measurements to the edge of the universe.

Suppose a galaxy shows a very large red shift, one so large that the galaxy would have to be moving from us at close to the speed of light. It can be found that when a galaxy is moving that fast, it must be 12.5 billion light-years away. It is at the edge of the universe that we can see.

If the galaxy had a slightly greater red shift, and its speed were equal to the speed of light, we would not be able to see it. The red shift would stretch out each wavelength so much, its energy would be reduced to zero. If the big bang occurred 15 billion years ago, we would expect the limit of the universe we can see to be close to 15 billion light-years. If so, there may be galaxies beyond the observable limit of 12.5 billion light-years. But they would have no meaning for us because there's no way we could be sure of their existence.

The actual size of the universe remains a mystery. What may lie beyond the part we can see is also a mystery.

NEUTRON STARS, PULSARS, AND LITTLE GREEN MEN

After a superstar explodes, what happens to the core that remains? That's a mystery that many people have wondered about. One of the first steps in solving any mystery is to reason out a solution. And this has been done.

What is a neutron star?

The core that is left after the explosion of a supernova is extremely dense. It contains a great deal of material and so has high mass and very strong gravitation. This pulls in other material, packing it tighter and tighter. The density becomes so great that electrons and protons are forced together to produce neutrons. The process continues until about 80 percent of the star has become neutrons. It is then called a neutron star. At the center, density may be 1 quadrillion. You remember the density of water is 1—1 gram per cubic centimeter. It means that a mass as great as that of the Sun, which is 1 400 000 kilometers across, would be compressed into a sphere only a few kilometers wide.

Most of a neutron star is made up of a gas so compressed

that it has become a liquid. The outer shell is made of atoms, free electrons, and protons. The atoms are mostly iron.

Gravitation of a neutron star must be unbelievably high—a hundred billion times stronger than the force of gravity here on Earth's surface.

Such were some of the ideas in a theory developed by many astronomers spanning many decades. It attempted to solve the mystery of what happens to the remains of the explosion of a superstar. For a long time little could be added, for no neutron star had been found. The ideas remained unproven.

What are pulsars?

By 1967 the theory could be checked. Radio astronomers had developed instruments that could make extremely accurate measurements. Using them, a strange star was observed.

At Cambridge University in England, astronomers were studying the sky to learn more about curious sources of radio energy—the quasars which will be discussed later. Most radio sources give out a steady hiss that is much like continuous static, or the sound you hear between stations on your radio. Jocelyn Bell, a student at Cambridge, noticed that there were unusual bursts from a new star, and that they occurred regularly. For several weeks the bursts remained a mystery. Three months later, on November 8, astronomers picked up the waves again, and found that they were quite remarkable.

The bursts lasted only about a hundredth of a second, and they arrived every 1.33730115 seconds. They were very regular. Thinking that there might be other sources like this one, astronomers searched the sky and found three more stars that behaved in a similar manner. They pulsed, so people called them pulsars. Since then several hundred have been located.

When optical astronomers looked at these stars, they discovered they were the same objects the astronomers had suspected were neutron stars. So, pulsars are neutron stars—small but massive stars that release tremendous amounts of energy.

What are Little Green Men?

When these strange stars were first discovered in 1967, and before pulsars were explained, someone jokingly suggested that the pulses were made by a far-off civilization. They were direction markers, it was said, that aliens used as they flitted from star to star. The radio pulses guided their journey, much the way ships and planes use radio and lighthouses here on Earth. For a time the pulsing stars were called LGMs. Some newspapers thought the astronomers were serious, so they printed stories saying that creatures beyond Earth had been discovered. There was considerable excitement around the world.

Why do pulsars pulse?

The pulses of pulsars are very short, often only a few fractions of a second in duration. This means the object that makes them has to be small, no larger than 3 000 kilometers across.

The reasoning goes like this. Suppose the Sun was pulsating. If suddenly sunlight were turned off, the center of the disk would darken first, because that's the part closest to us. Then the darkness would spread to the edge of the Sun. Because of the size of the Sun, about two seconds would be needed for the darkness to travel some 700 000 000 kilometers—from the center of the disk to the edge. That's much longer than the pulses of a pulsar, so an object the size of the Sun could not be a pulsar. It has to be only about 3 000 kilometers across to pulsate as fast as the rates that have been observed.

A major mystery of pulsars was how even such small objects could pulsate so rapidly—in some cases thirty times in one second. Perhaps it is because they do not change from light to dark in the manner described above. More likely, they rotate very fast. Our Sun rotates, taking about a month to complete one turn. Theory says a massive object like a pulsar could rotate much faster, a thousand times a second, without flying apart. So it could certainly rotate more slowly—at thirty times a second, or ten or twenty.

Radio astronomers continued to search the skies for pulsars, and optical astronomers tried to see them. The places they looked were those locations where star explosions had occurred. That's where stellar cores should be found.

What is the Crab Nebula?

In the year 1054, Chinese astronomers saw a brilliant object in the constellation Taurus, the bull. It was so bright that the object could be seen in the daytime.

Much later, astronomers used telescopes to look in the direction indicated by the Chinese. They found a large formation of gases, a nebula as it is called. The sprawling shape reminded them of a crab, so the formation has been called the Crab Nebula. The gases were moving at high speeds away from a central core. They were speeding away because they were remnants of a massive explosion, which could have been the bright light the Chinese had seen centuries earlier. If that is so, the gases should still be expanding. And they are—at speeds of about 1 000 kilometers a second.

In the central region of the nebula, there is a star—a neutron star—that pulsates thirty times a second. It rotates rapidly and gives off tremendous amounts of energy. The same star had been picked out some twenty-five years earlier as a possible remnant of a star explosion. But at that time, in the 1930s, its pulses could not be noticed because they were too fast. The light blended together since the eye cannot see separations between the pulses. But radio telescopes later separated such rapid changes.

The star pulses, or seems to, because it rotates on an axis that is at right angles to our lines of sight. The north magnetic pole of the star swings alternately toward us and away from us. Charged particles are released from the polar region. Every time the pole swings toward us—thirty times a second—we receive a pulse of radio waves from these particles. They are contained in

The gases making the Crab Nebula are remnants of a star explosion that was seen in A.D. 1054. The gases continue to expand at 1 000 kilometers per second. Toward the center is a neutron star, the central object that remained after the explosion.   PALOMAR OBSERVATORY PHOTOGRAPH

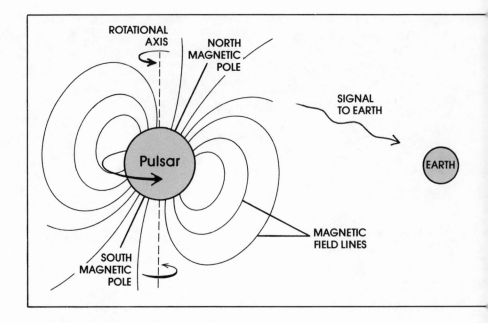

A pulsar does not pulse. It spins rapidly. Each time a magnetic pole is toward the Earth, we receive energy from the star. The radio waves therefore seem to pulse with a very rapid period. The Crab pulsar "pulsates" thirty times per second.

a tight beam that sweeps by Earth, much as the beam from a lighthouse sweeps by a ship. The pulsar is not actually pulsing; it is rotating rapidly and so gives the impression of rapid pulsation.

If this theory is correct, the pulsar should be slowing down; and it is. This happens because of drag produced by the particles moving along and away from the magnetic lines of the star.

Attempts to explain in greater detail the behavior of pulsars have occupied astronomers since their discovery in 1967. Much about them remains a mystery that challenges many people.

BLACK HOLES

What is a black hole?

This chapter will investigate that question. At the start, it can be said that black holes are major mysteries of the universe. They don't give off light, so you can't see them. They don't give off X rays, so you can't "see" them that way. Neither do they give off radio waves, so radio telescopes can't "see" them either. How, then do you know that they exist?

Black holes seem to be formations much smaller than the Sun but containing at least three or four times more material than is in the Sun, and often much more than that. The material is incredibly dense, intensely packed together. If the mass of the Earth were to be compressed into a black hole, the object would be no larger than a baseball. A single teaspoonful of a black hole would weigh billions of tons.

Black holes are so mysterious that many people find it hard to believe that they exist. Others are sure they do, and astronomers believe they have located at least one of them. It is in Cygnus, the swan, a constellation we can see in our summer skies. It is called Cygnus X-1.

How were black holes discovered?

In 1971 an American satellite was launched from Kenya in Africa. It was called Uhuru, a Swahili word meaning freedom. Uhuru was an X-ray satellite. It was sent above the atmosphere, and so was able to receive X rays. In particular, it was looking for X-ray stars, stars that were releasing large amounts of high-energy radiation—hard X rays, as it is called.

Uhuru was to discover over a hundred X-ray stars. In particular it found a source in Cygnus, the swan, that was uneven—the X rays coming from the vicinity varied in intensity. First they were strong, then they were weak, and the time between peaks was very regular. Uhuru zeroed in on the region and gave observers the exact location of the source.

Radio astronomers on Earth turned their instruments to that location, and they found that radio waves were coming from the area. When optical astronomers looked at the region, they discovered a very large, hot, blue star at the same spot.

Careful study of the light coming from the star showed that the light kept changing. Alternately it became bluer and then redder. There was a blue shift as the star moved toward us, then a red shift as the star moved away. Stars cannot move toward and away from us along a straight line. In order for the star to show these shifts, it had to be going around something very massive. There were two objects at the location—the visible star and another object that could not be seen. To explain it, astronomers imagined a model of the formation.

The invisible companion, whatever it was, was pulling the blue star apart. Gases were leaving the large star and going into the companion. As the gases streamed from one mass to the other, they became very hot, so hot that they gave off X rays—the X rays that Uhuru had picked up. The X rays were very strong, indicating that tremendous amounts of gases were being swallowed by the companion. The X rays were given off just before the gases disappeared into the massive object. So, although it was invisible, the X rays revealed that the large blue star must have a massive companion.

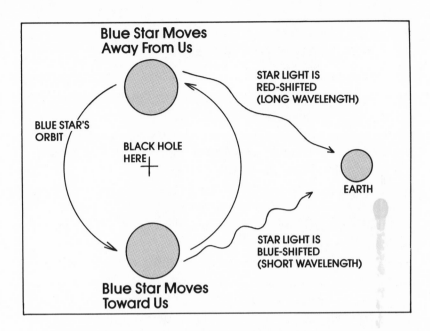

**Blue Star Moves
Away From Us**

STAR LIGHT IS
RED-SHIFTED
(LONG WAVELENGTH)

BLUE STAR'S
ORBIT

BLACK HOLE
HERE

EARTH

STAR LIGHT IS
BLUE-SHIFTED
(SHORT WAVELENGTH)

**Blue Star Moves
Toward Us**

Cygnus X-1 appears to be a two-star system; one huge and blue, the other a black hole. As the large star moves toward us, its light shows a blue shift; as it moves away, there is a red shift.

The strong gravitation of the black hole pulls gases from the large blue star. The gases are swallowed up and disappear. The black hole gives off no radiation—no radio waves, X rays, or light by which we can "see" it—hence the name black hole.

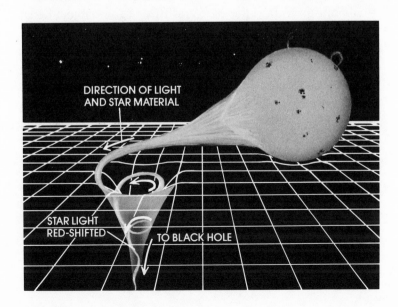

DIRECTION OF LIGHT
AND STAR MATERIAL

STAR LIGHT
RED-SHIFTED

TO BLACK HOLE

How are black holes different from neutron stars?

In order for a neutron star to form, the star that gives birth to it must have a certain mass—it must contain a certain amount of material. That's because neutron stars are produced by contraction. And the amount of contraction depends upon gravitation. The greater the mass, the greater the gravitation. The Sun could never become a neutron star, for it does not contain enough material. Gravitation could not be great enough to pack the material close enough to reach the density of a neutron star. Astronomers believe that in order for a star to become a neutron star it must have a mass two or three times greater than that of the Sun, but no more than that.

Should a star be more massive, it would have enough gravitation to pack itself even tighter. Density would become much higher than the density of a neutron star. The mass would become a black hole.

All the observations of Cygnus X-1 could happen if the companion of the blue star was a neutron star—one that was not bright enough to be seen optically. But perhaps the companion was much more massive than a neutron star.

Back in the 1930s, astronomers had speculated that there might be supermassive stars. But there was no instrument sensitive enough to find them. Could this Cygnus object be one of those supermassive stars? To find out, astronomers had to figure out its mass.

They knew that the visible blue star was massive. To produce the light and other energy that it did, the star had to have a mass perhaps thirty times that of the Sun. By measuring the blue and red shifts of that star, astronomers could find how fast the star was going around its companion. Knowing that, it is possible to compute how large the companion had to be. It was

Stars generate energy by nuclear fusion reactions. During billions of years, they go through stages of heating and cooling, expanding and contracting—and exploding. In their final stages, those more massive than the Sun enter a collapsing stage which does not stop. The star collapses into zero volume and infinite density; it becomes a black hole. ▶

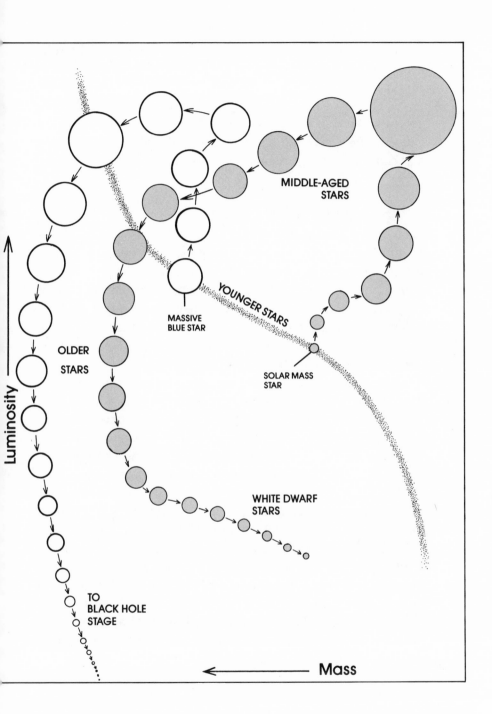

MIDDLE-AGED
STARS

YOUNGER STARS

MASSIVE
BLUE STAR

OLDER
STARS

SOLAR MASS
STAR

WHITE DWARF
STARS

TO
BLACK HOLE
STAGE

Luminosity

Mass

found that the mass of the object was close to eight solar masses. That was too massive for it to be a neutron star. The object had to be a black hole.

We cannot see the black hole, for we do not receive light or any other kind of radiation from it But we know it is there because of the X rays and also because of the effects on the blue star caused by the companion's powerful gravitation.

How do black holes form?

Black holes may be the end products of massive stars. While fusion reactions are occurring in stars, the energy produced pushes outward. Gravitation of the star tends to pull the star inward, but the outward pressure keeps this from happening. When nuclear fuel is used up, pressure outward drops. Pressure inward is still high, and it causes the star to collapse. Should the star have a mass two or three times that of the Sun, the mass becomes a neutron star.

When the mass is greater than this, the collapse continues beyond the neutron-star stage. The mass becomes a black hole. Volume of the mass continues to shrink until it becomes zero. And density of the mass continues until it becomes infinite. Both conditions are deep mysteries, for no one can conceive zero volume and infinite density. Escape velocity, the speed needed for anything to escape from a massive object, becomes greater than the speed of light. That is why we cannot see a black hole—nothing can escape from it.

Do black holes really exist? They seem to. But there are many mysteries about them. Some say black holes are common in the universe, and tremendous amounts of matter are concentrated in them. They say there's enough mass in black holes to cause the universe to collapse. Perhaps so, but that's another and a different mystery. It's one that will be considered after first taking a look at quasars—quasi (much like) stars.

7 PERPLEXING QUASARS

Are quasars at the edge of the universe?

Could there be something that looks like a star, yet gives off more energy than 5 trillion Suns? Could there be an object that in one second gives off energy equal to Earth's needs for a billion years? The answer is yes; they are quasi-stellar objects—quasars for short. And therein lies one of the most puzzling of all mysteries.

Quasars are starlike objects; in a telescope they appear like stars. They cannot be seen without a telescope, because they are too dim. Actually, however, they are extremely bright. *Luminous* would be a better word, for their brightness might not be in visible light but in radio waves, or ultraviolet or X rays. They appear dim only because of their great distance from us. Some seem to be at the edge of the universe.

When first observed in the 1960s, quasars could not be identified because the spectrum they produced was not familiar. After considerable study, the reason for the different spectrum was found. The lines were shifted much farther toward the red than any spectra that had ever been observed.

As has been pointed out, there are relationships between red shifts, distance, and speed. Red shifts indicate there are quasars that are moving 240 000 kilometers a second, 80 percent of the speed of light. Also, quasars appear to be farther away than any other object in the universe. Some are 12 or so billion light-years away. That means we are seeing the objects as they were 12 billion years ago. They are the "youngest" of all objects—formations that were created only a few billion years after the big bang. We are looking backward in time. Indeed, quasars may have gone out of existence long ago. But because of their great distance we would not know that had happened.

How do quasars produce energy?

If quasars are so distant, they must produce gigantic amounts of energy for any of it to reach us. It has been estimated that some of them generate in one second more energy than our Sun produces in 3 000 000 years. Quasars appear like stars, yet they produce energy equal to that generated by dozens of entire galaxies. They do this periodically, sometimes every few days or weeks, or perhaps every year.

The period of the variations in energy output provides clues to the size of the objects. If the period is one year, then light would take a year to travel across the object, and so the object could be no more than a light-year across. That is remarkably small for an object that produces so much energy. Galaxies give off only a fraction as much energy, yet they are much larger. Our galaxy, for example, is 100 000 light-years across.

What kind of object could be so small and still generate so much energy? That's a major mystery; however, many solutions have been suggested. The energy seems to come from electrons spinning at high speed in a magnetic field. (It is called synchrotron radiation, a kind of radiation first noticed here on Earth in a man-made subatomic reactor called a synchrotron.) At regular periods, the supply of electrons is renewed in some fashion.

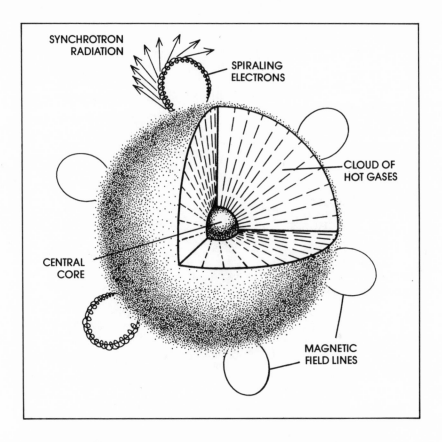

SYNCHROTRON
RADIATION

SPIRALING
ELECTRONS

CLOUD OF
HOT GASES

CENTRAL
CORE

MAGNETIC
FIELD LINES

A model of a quasar. The diameter of the cloud is several light-years, and its mass is about 10 000 000 times greater than the mass of the Sun. Energy is generated in the small central core. Electrons spiral along magnetic lines of force, producing radiation.

The outer part of the quasar appears to trap the energy of the electrons and transform it into infrared, ultraviolet, X-ray, and visible light and radio energy.

But how are the electrons replenished? No one has the answer to that mystery. However, it's possible that the quasar is spinning rapidly, and it is very massive. A spinning object packs a lot of energy which very likely is transmitted in some manner to the electrons.

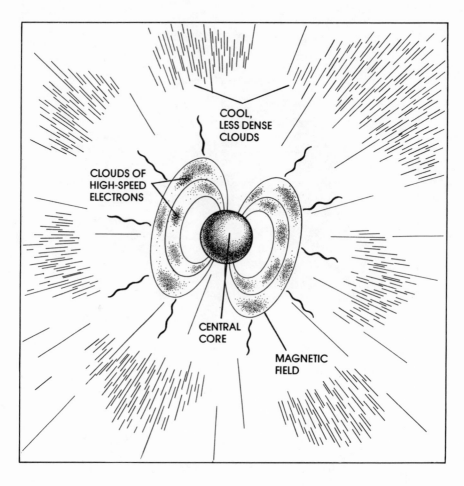

COOL,
LESS DENSE
CLOUDS

CLOUDS OF
HIGH-SPEED
ELECTRONS

CENTRAL
CORE

MAGNETIC
FIELD

Energy is generated in a superdense central core. The spiraling electrons produce synchrotron radiation. Cooler outer clouds absorb some of the energy generated.

Are quasars super black holes?

Quasars appear to have been formed at an early stage in the evolution of the universe. Perhaps they are usual first stages in the development of galaxies. They may be young super galaxies that eventually quiet down to become normal galaxies such as our own. If so, galaxies should have intensely dense cores where mass is concentrated. These would not be unlike massive black holes, as some people have suggested.

In fact, quasars themselves may be super black holes. At the beginning, a young galaxy may have had at its core a black hole a 100 000 000 times the mass of the Sun. Stars that formed in that core region may have been supermassive—a hundred and more times the mass of the Sun. That means they would have burned out rapidly, taking only a few million years to reach the black hole stage. Dozens of stars would have evolved similarly,

Quasars move at high velocity and have the greatest red shift of any known object in the universe. Notice that they merge into radio galaxies (those that give off excessive radio energy), and these merge into normal galaxies. Perhaps quasars are a young stage in the formation of these "normal" galaxies.

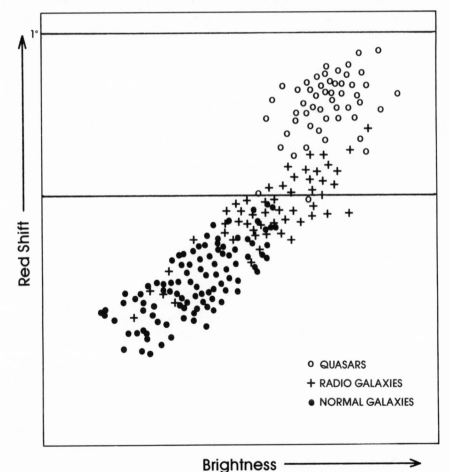

and the black holes created may have combined to feed the super formation.

Gravitation of such a super black hole would have been stupendous, great enough to pull apart nearby stars. As gases from these stars spiraled into the hole, energy would have been set free and might have become the source of energy for the quasar. Each year a solar mass of matter might be swallowed up. The activity of the quasar would decrease rapidly as nearby material was consumed. Energy generation would diminish and the quasar would evolve into a quiet galaxy.

If such ideas are the correct solutions to quasar mysteries, astronomers should "see" X rays coming from a tight core in galaxies. Or, there might be intense light at the core caused by concentrations of stars. The stars at the core should also be revolving rapidly, for the same reason that a skater spins rapidly when mass is concentrated—the skater's arms are wrapped tightly around the body. Such observations have been made in some cases, so believers in these theories are encouraged. However, many more observations are needed before the theory becomes the solution.

Quasars remain a fascinating mystery of astronomy, not only because we think we are seeing the universe as it was billions of years ago, but also because astronomers are challenged to find logical explanations for the unbelievable amounts of energy that are generated. George Gamow, a scientist who was concerned about such things, wrote this rhyme some time ago. It expresses very well the feelings of a good many quasar searchers.

Twinkle, twinkle, quasi-star
Biggest puzzle from afar.
How unlike the other ones,
Brighter than a million suns.
Twinkle, twinkle, quasi-star,
How I wonder what you are.

Quasars, neutron stars, black holes are puzzling parts of the universe. Equally puzzling is the mystery of its future. How long will the universe last? Will it ever end? If it should end, how would it happen?

END OF THE UNIVERSE

Will the universe ever end?

The universe is expanding. Wherever one looks, the astronomer sees a red shift. The light sources are moving away, so there's no question about the present direction of the universe. But there's considerable mystery about its future. Will the universe expand forever? Is the universe open? Many people believe so.

Others contend that the universe pulsates, which means we live in a closed universe. Right now, they argue, the universe is expanding. But that is only temporary. Eventually, these pulsators say, the expansion will slow down. It will come to a stop, and then instead of expanding, the universe will collapse. The process will take billions of years. They may be right. There are no definite answers, so the future of the universe continues to be a mystery.

Gravity holds the answer. Very simply, if gravity is not strong enough, the universe will go on expanding forever. There will be no end to time. On the other hand, should gravity be sufficient, it will slow down the expansion, bring it to a stop, and cause an eventual collapse. So the mystery to be explored is the gravity of the universe. How do we determine what it is?

What is the mass of the universe?

It is known that gravity depends upon two factors—one is mass and the other is distance. The larger the mass, the greater the gravitational attraction. And the closer the masses are together, the greater is the attraction between them.

When one knows the size of the universe and its average density, mass can be determined. Astronomers concerned about such matters have found that in order for gravity to be great enough to stop expansion, density of the universe must be 5×10^{-30} grams per cubic centimeter. (The -30 means 30 places to the right of the decimal point. For example, $1 \times 10^{-1} = .1$; $1 \times 10^{-5} = .00001$; so 5×10^{-30} is decimal point followed by 29 zeros and a 5.) In a universe of this density, there would be one particle in every cubic meter.

Are there that many? That's the problem. One must find the average density of the universe. If it turns out to be at least 5×10^{-30} g / cc, the universe is closed. Should the density be less than that, there will not be great enough gravitation to stop expansion, and expansion will continue forever.

First steps toward solving the mystery were to count visible galaxies as well as those believed to exist, and then to determine their mass and density—a difficult task, if not impossible. However, in theory at least, it has been done. Density turns out to be alarmingly low, only about 2 percent of the density needed to close the universe.

But not all the matter of the universe is in galaxies. There are star clusters of hundreds of thousands of stars and there is intergalactic gas—gas between galaxies. What is the density of this matter? It's extremely difficult to say, just as it is almost impossible to get an accurate number for the density of the galaxies. However, one can reach conclusions on the basis of reasoning. All the matter believed to exist outside of galaxies adds up to a very small fraction of that in galaxies. So far as has been determined, all parts of the universe produce a density that is only 10 percent or so of that needed to make a closed universe. So, from that standpoint at least, the universe is open—it will continue to expand.

Is deuterium a key to the universe?

There are other considerations that support the theory that the universe is open. One is the amount of deuterium in the universe. As was mentioned earlier, during the first few moments after the big bang, hydrogen was created, along with helium and deuterium. (Deuterium is a hydrogen atom with two mass particles, a proton and a neutron, in its nucleus.) The amount of original deuterium that has survived to the present day depends upon the density of the universe.

If the universe should be extremely dense, there would be very little, if any, deuterium. That's because, above a certain density, deuterium nuclei in the early universe would have combined with others to produce helium. If the universe should not be very dense, deuterium would have remained because at low density, deuterium nuclei would not have become helium.

Astronomers searched for deuterium, but until the early 1970s none had been found. It is identified by the ultraviolet light it gives off, and most ultraviolet light cannot penetrate our atmosphere. In the 1970s, ultraviolet satellites were sent above the atmosphere. They searched the sky and found that deuterium was there. Therefore, the density in the very early universe could not have been great enough to convert all the deuterium into helium. The presence of deuterium implies that the density of the universe has never been great enough to cause collapse. So, the deuterium yardstick is another indication that the universe is open.

How will the universe end?

Should the universe be open, as it seems to be, what will be its future? It will continue to expand. As the universe expands, it will cool. The background temperature of three degrees will eventually drop to zero.

Stars within galaxies will complete their life histories. Those less massive will cool down. Life, which is dependent upon star energy, will disappear from Earth and from any other location in the universe where it might have existed. Massive stars will become neutron stars, pulsars. Whatever radiation they give off will be produced by synchrotron radiation and not by high temperatures. More massive stars will become black holes. And, as many suggest, entire galaxies may become supermassive black holes. The evolution of the universe will have come to an end.

Many people find it hard to accept that there can be an end of any kind. They contend that change of some sort must go on forever. They believe that astronomers really don't know the density of the universe. They say there are factors that will eventually reveal that there is much more matter in the universe than is now believed to exist. They contend that because of this increased matter, the universe must be closed. Maybe so.

According to this theory, gravitation is sufficient to slow down expansion. The universe will stop expanding. Then, after billions of years, there will be movement in the opposite direction. Galaxies that have been moving farther apart will move closer together. Because of unequal gravitation on them, the galaxies will be pulled apart. As packing of the part continues, temperatures will break the galaxies down into individual atoms, and eventually into atomic particles—protons, neutrons, and electrons.

Density will increase millions of times and temperatures will soar to billions of degrees. Another cosmic egg will be created and, like the original, forces within it will not be contained. There will be a second big bang. And with that, the series of events you read about earlier will recur.

The universe may be pulsating. At the present time it is expanding. But perhaps this expansion is only a stage in the history of the universe. Expansion may stop, to be followed by

Open Universe

Closed Universe

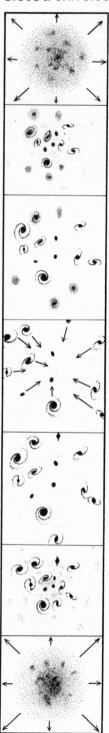

contraction. And this might be followed by another expansion. And so on forever.

That may be. Presently no one can say one way or the other. It is a major mystery, and one that will continue to be a puzzle. Like many other mysteries, it may never be solved.

But in the years ahead, no doubt new instruments will be developed to explore the universe more deeply. Our knowledge will increase. We may have just begun to learn how to use our minds to accurately define black holes, pulsars, quasars, and all the other phenomena of the universe. We may someday learn exactly what our history has been, and what may eventually happen to the universe.

If the universe is open (low density), all parts of it will continue to expand forever. Stars and galaxies will burn out; the universe will run down and end. Should the universe be closed (high density), expansion will come to an end. The universe will contract, perhaps into a cosmic egg, and there will be another big bang. The universe will "pulsate," and so will never end.

FURTHER READING

Branley, Franklyn M. *Black Holes, White Dwarfs, and Super-stars.* New York: T. Y. Crowell, 1976.

———. *Halley: Comet 1986.* New York: Lodestar Books, Dutton, 1983.

———. *Jupiter: King of the Gods, Giant of the Planets.* New York: Lodestar Books, Dutton, 1981.

Kippenhahn, Rudolf. *100 Billion Suns.* New York: Basic Books, 1983.

Morrison, Phillip, et al. *Powers of Ten: The Relative State of Things in the Universe.* Scientific American Library Set. San Francisco: W. H. Freeman, 1982.

Shipman, Harry L. *Black Holes, Quasars, and the Universe.* Boston: Houghton Mifflin, 1976.

Trefil, James. *The Moment of Creation.* New York: Scribner, 1983.

Weinberg, Steven. *The First Three Minutes: A Modern View of the Origin of the Universe.* New York: Basic Books, 1976.

INDEX

black holes, 43–48
  density of, 43, 46, *46,* 48
  discovery of, 44
  escape velocity from, 48
  gravitation of, *45,* 46
  light and, 43, 48
  mass of, 43, 48
  neutron stars vs., 46–48
  quasars as, 53–54
  weight of, 43
  zero volume of, *46,* 48
blue shift, 34

calcium, red shifts of, *35*
Cambridge University, 38
carbon, creation of, 13
carbon stars, 21
Celsius scale, 16
Centaurus, *29*
Chinese astronomers, 40
Corona Borealis, *31*
cosmic background radiation, 16
cosmic egg, 11
  density of, 12
  explosion of, 12
  galaxies formed from outer portion of, 15
  second creation of, 59, *61*
Crab Nebula, 40–42
  expansion of, 40, *41*
Cygnus X-1, 43–48
  discovery of, 44
  mass of, 46–48
  as two-star system, 44, *45*
  X rays emitted by, 44, *45*

density:
  in big bang, 12
  of black holes, 43, 46, *46,* 48

density, *cont'd.*
  of cosmic egg, 12
  definition of, 12
  of galaxies, 57–58
  infinite, *46,* 48
  of neutron stars, 46
  of universe, 57–60
  of water, 12
deuterium, *19*
  density of universe measured by, 58
deuterons:
  formation of, 12, 19, *19*
  helium formed by, 12–13, 20
  in hydrogen, 19

Earth:
  distance of Sun from, 32
  formation of, 14
  stardust on, 25
electrons:
  in deuterium, *19*
  in hydrogen, 18
  neutrons formed by, 20
  quasars and, 50–51
  in tritium, *19*
elements:
  in fusion reactions, 20–22, *24*
  heavier, rareness of, 20–21
  "old" vs. "young," 13
  spectrographs of, 33–34
  in star explosions, 13, 22–25
  unstable, 25
energy:
  in big bang, 11–13, 18
  in fusion reactions, 20
  object size and output of, 50
  from quasars, 49, 50–51

energy, *cont'd.*
  radio, 38-39, 40, 43, 49
  from star explosions, 21,
    24-25
  at time zero, 11
Eridanus, *28*
escape velocity, 48
expansion of universe, 56-60
  beginning of, 12
  closed universe vs., 56,
    57-58
  density and, 57-60
  galaxies formed in, 13-14
  gravity and, 13, 56, 59-61
  inflationary universe and,
    14-15
  red shift in, 15

"fossil light," 32
fusion reactions, 20-25
  after big bang, 20
  elements formed in, 20-22,
    24
  energy released in, 20
  iron and, 22, 23
  outward pressure during,
    48
  self-sustaining, 20
  in stars, 20-22, 23, 46

galaxies, 4
  clusters of, *31*
  cores of, 52-53
  creation of, 13-14
  density of, 57-58
  discovery of, 4, *26*
  distances between, 9, 36
  flattened shapes of, 14
  gases in, 14
  in "local group," 32
  most distant, 15, 32

galaxies, *cont'd.*
  number of, 9
  photographs of, *26, 28-31*
  quasars and formation of,
    52-53
  red shifts of, 15, 32-36
  spinning of, 14
  spiral, *28, 30*
Galileo, 4
Gamow, George, 11, 54
gases:
  in neutron stars, 37-38
  stars and planets formed
    from, 14
  stellar formations of, 40
gravitation:
  of black holes, *45*, 46
  expansion of universe and,
    13, 56, 59-61
  mass and, 57-58
  of neutron stars, 37, 38, 46,
    48
  of stars, 20-21, 48
  of universe, 13, 53, 59-61
Great Nebula, 7

half-life, 25
helium, 13
  age of, 18
  atomic number of, 10,
    12-13
  creation of, 12-13, 20
  in mass of universe, 13
  nucleus of, 10, *19*
  in star explosions, 24
hydrogen, 10, 18-20
  age of, 18
  atomic number of, 10, 18
  creation of, 11-12
  in mass of universe, 13
  nucleus of, 10, 19, *19*

ABOUT THE AUTHOR

Franklyn M. Branley is the popular author of over one hundred books for young people on astronomy and other sciences, including *Halley: Comet 1986*, *Space Colony*, and *Jupiter*. He is Astronomer Emeritus and former chairman of The American Museum–Hayden Planetarium. He and his wife live in Sag Harbor, New York.

ABOUT THE ILLUSTRATOR

Sally J. Bensusen, illustrator of *Halley: Comet 1986*, has done work for the Smithsonian as well as for many astronomy magazines. She lives in Lanham, Maryland.